"The Secrets In The Woods"

Publisher and Author

Donald James Quinney

Book Design

Donald James Quinney

And

Ryne Sullivans

"Introduction."

First I want to thank Ms.Lackey, for open her mind up, for this book to evolve. Sometimes in life, you do not know what God has for you in this life, but you must use your talent in a way that strengthens your soul. This is a thriller, about a woman, that lost her husband in the deep forest, three years ago.

Still caught up in the illusion, that her husband is still alive, she can't let go. She has been coming to the edge of the forest for three years, wanting to enter to look for her husband

When she went to the authority to find him, they told her, people do go missing, and are never found in that forest. So she decided to finally, to go and find any clue of what may have happened to him. Two days later, she came to the entrance of the place, and she enters. She looks with fear in her eyes, but with mix emotions in her heart, she approaches, to go into the unknown.

The humid air made her sticky; she pulled her hair up into a ponytail to keep it from getting murky.

Looking out a cardinal caught her eye, hopping along the forest floor looking for bugs. A slight breeze whispered through the trees bringing the scent of pine and dirt.

Sighing, she stood from the hard bench she'd been sitting on and slowly made her way towards the path. As she walked deeper into the woods, the outside world melted away, and nature came to life.

Little squeaks from squirrels and the chirping of birds were almost overwhelming as she turned around the bend in the path. "If only he were here to see this," she thought to herself.

Letting the thought go she trudged on allowing the forest take her in its simplicity and beauty. A loud

Crack! made her jump and her heart pump faster, spinning around she saw nothing but trees.

"Maybe it was a deer," she said aloud trying to convince herself. Being afraid of the situation she is in, but need the answer, she is looking for.

Turning she continued on the path the twigs crunching underfoot making her steps loud to her.

Beautiful birches and towering pines decorated the sides of the path, little chipmunks and birds scurried and hopped about trying to figure out what she was.

"It's so peaceful," she thought. Smiling and deeply inhaling the scent of pine and other beautiful things she walked deeper and deeper into the woods.

 The sounds of cars and people completely muted for over an hour now. Something big rustled the bushes ahead making her stop. A small fawn hopped and ran back whence it came. She grinned and started walking again whistling a tune she heard.

On the radio last night. It was a while before she noticed the sounds of the forest were gone making the trees surrounding her eerie and quiet.

Perspiring a little, she walked a little faster her mouth dry from the trepidation she felt. "Stop being silly you just scared off all the animals," she thought. Digging into her pack, she took a swig of water trying to.

Quench the terrible thirst that seized her. With a spasm of fear she suddenly realized she'd lost the path, whipping around she searched the ground looking.

 As a sense of dread filled her, she knew she was lost and very, very deep in the woods. Trying not to cry she started walking again praying she'd find the path still. "He wouldn't have gotten lost," she thought, mentally smacking herself on the forehead. Slowing her breathing she

Stopped trying to remember which was the moss grew on trees or whatever that saying was. A loud clap of thunder made her blood crawl and the hairs on her neck stand up.

Cursing at herself for not bringing a com-
pass or umbrella, though there weren't
any clouds when she

started walking, she searched her pack for
a parka or anything that'd keep the rain
off. Finding nothing

she searched around for an excellent tree
to shelter under for when it started rain-
ing, locating a big oak tree

She made herself a makeshift tent with big
branches from a few pines a couple of
yards away. It was good

Timing as she sat under her shelter when the rain started coming down in torrents making it

impossible to see two yards ahead of her. She curled up on the forest floor using her pack as a pillow.

And fell asleep the stress from getting lost taking her. Amanda started to dream about her lover and filling that she is getting closer to him.

Waking disoriented and forgetting about

getting lost, she panicked a little wonder-
ing where she was it

all came back in one big ugly flash. Sitting
up she looked around her; the rain has
stopped which meant

she could start walking again to find the
path. Reaching for her pack, she gasped as
her hand held up an

empty and ruined pack. All of her food,
water, toiletries were gone, and her pack
was ripped and torn as if it was an animal.

Desolate and dirty from sleeping on the ground she stood and walked aimlessly. She thought about an

evening three years ago, twirling and swaying to the music softly playing on her husband's radio. How

happy she'd been molding her body to his and smelling the aftershave he used that drove her crazy. A

distant thunder roll brought her back to the present; lost, hot, and thirsty, she started looking from a

Creek for some water. Her heart ached
thinking of her husband; he'd been missing for five years now

which was why she was lost in these
woods now, the very forest that he'd gone
missing in.

 A cold sweat formed on the nape of her neck fear trying to take hold, swallowing her anxiety she

continued her search for a creek. Her attention was brought back to the eerie silence of the forest, not

a bird or squirrel in sight nor sound.

 "Oh Stephen I wish you were here," she said aloud with a pain filled voice.

She heard a whisper behind her and swiv-
eled around finding no one. Praying to
God, she turned back

around and started walking again. Stomach growling she looked around for some.

Berries, anything to

eat for she felt like she was starving. Suddenly she realized that strewn across the
ground mixed with

the dirt was the food she'd had in her
pack; she cried because she knew she
couldn't eat it, she

examined it closer her eyes widening. "This had to have been a person, an animal would've eaten the food," she thought her cold sweat

coming back. Walking slightly faster she had her eyes peeled for any motion and listened for anything.

That didn't sound like her footsteps. Night
came quick like a ninja catching her una-
wares. Fear took hold

in the quiet black forest; she saw the roots
of a tree making a kind of hide hole and
crawled in. She

curled up trying to stave off the tears not
wanting to waste what moisture she had in
her body.

 Waking to a beam of sunlight on her face and a terrible thirst she stretched and climbed out of the roots.

It seemed to be about 7 am, but without a watch, she had no clue as to what time it was. Hopelessness

set in as she looked around recognizing nothing, how was she ever getting out of these woods she

Thought, Her mind took her back a few years to her first wedding anniversary. Stephen had woken her

up with kisses starting at the hollow of her neck and trailing down.

As she awoke to pleasure and the subtle scent of him she moaned and arched her back to get him

better access. After she had climaxed, he kissed her smiling that dazzling smile of his that never failed to

cheer her up. Grinning from ear to ear she told him the happy anniversary and caressed his jawline.

"I love you so much sweetie," he said entwining his body with hers. She murmured I love you too,

snuggling closer. As the day progressed, both had surprised each other with sweet gestures and small

gifts.

"Stephen!!," she yelled in anguish and fear at being so lost in these woods.

Another whisper made her almost soil herself in terror; it was his voice. Slowly turning around she saw a

figure flit through the trees disappearing before getting a good look at whoever it was. Heart pounding

She looked about for something to use as a weapon, if whoever it was to attacked her no one would hear

her screams. Finding nothing she nervously looked around and started walking again. Remembering

she'd packed a knife in her pack, she almost broke down when she realized that
among the destroyed

food she did not see her knife. Hearing
something in the trees, she started walking faster then broke into a run branches
slapping her.

Face and almost tripping over roots and
rocks. Stumbling over a large root, she fell
tumbling down an

incline. Standing up she brushed herself
off looking around and spotted a creek.
Gratefully she ran to it.

Gulping down mouthfuls using her hands. After she had her fill she stood and started following the creek
not wanting to part from it.

 "Hello," a deep timber voice said, making her shout in fright.

 Spinning around she saw a man in blue jeans and a button-down; he looks like a wild man and his hair

was long and black as a raven. He seemed like a mountain man; his skin was tight from the weather and

his eyes wide and curious. He watched her scrutiny with an unreadable expression on his face, she looked at his belt and realized he had a gun. Nervous she licked moisture onto her lips and tried to think

of something to say that wouldn't make it seem like she'd noticed his weapon.

"What are you doing so deep in these woods?" he asked startling her.

"I...I'm lost," she stammered.

She rubbed her clammy hands on her pants trying not to show how nervous she was. The hairs on the back of her neck stood up looking at the man, something didn't seem right with him, but she couldn't pinpoint what. "My name is Catrina, I was hiking yesterday and got lost," she told the man.

"I am Ravan, I live not far from here," he said introducing himself.

"You look like you haven't eaten would you care to join me for lunch?" he asked with a smile trying to set her at ease only making her more nervous.

Her stomach was betraying her growling loudly she accepted and followed him in the opposite direction she'd been heading. They walked for a long time in silence her feet aching, along the way the

Creek thinned out and disappeared the farther they walked. She was about to ask how much further when she spotted a cabin up ahead.

"That is my home," Ravan said, "I live alone, I don't like the pollution that man has made. Everything I have and everything I make is natural and comes from the forest."

"It's beautiful," she exclaimed taking in every detail. There was a fire pit out front; the cabin was made of oak nothing adorned the porch except a few dream catchers here and there made with feathers and twine the man no doubt made himself. A trough sat to the side of the house filled with fresh water.

"Do you have a horse?" Catrina asked.

"No," he said with no more explanation.

They walked into his house in silence, he

Pointed to a chair for her to sit and proceeded to make them food. A delicious aroma of berries, pancakes, and eggs filled the house making her stomach growl loudly.

Still quiet he served her food to her and ate without saying a word. She let her gaze roam around his house, simple but quaint he had one other chair than the ones they were sitting on. A picture of an

old Native American in full dress hung above his fireplace.

Flashback a few years she thought of the day that her husband disappeared. They'd woken up together on a Saturday the day started beautifully, making breakfast Catrina had lazed about dancing to the music on Stephen's radio both laughing and enjoying their rare day off.

Stephen had planned a hike with a friend from work to take some photographs of a rare bird they'd been searching for.

Catrina had planned to continue her book and packed his backpack the day before with dry goods, water, etc..

As the day progressed and Stephen had already left she finally finished her book feeling accomplished.   Nearing 7 p.m. she grew worried when Stephen didn't come home, trying not to panic she texted and called him and his friends repeatedly phone as it got later and later.

At 10 p.m. she called the police to report her husband missing, they told her he was an adult, and they couldn't do anything until he'd been missing for at least 24 hours.

Waiting sleeplessly she called the police the next day at the same time and filed the report.

  Accompanying them on their search, she grasped at her sanity the longer they were in the woods looking for her husband and his friend. At 4 a.m. the policeman had to call it a night, everyone was tired, and there was no light to search by.

They promised they'd continue their search later that afternoon and pray for the best. She stayed a few more hours at the edge of the path staring into the trees sobbing and calling for Stephen until her throat hurt, and her voice was hoarse.

A loud cough brought her back to the present, and she realized Raven had been talking to her.

"I apologize my mind was elsewhere, what were you saying?" she asked pushing the painful thoughts into the back of her mind.

"I asked what made you take a hike in these woods," he asked politely.

"This was the forest my husband went missing in 3 years ago; I come here every year on the day he went missing," she told him sadness saturating her voice.

He sat there musing over it not comment, she didn't meet his gaze and looked around her soaking up the details of his small home. There was a door that most likely led to his bedroom off to the left of the fireplace.

A small handmade table stood near the kitchen; she could see the stove from where she sat, no refrigerator adorned his kitchen which she thought was odd. Listening carefully she could hear a few chickens out back which explained the eggs.

"Thank you for your kindness and the food, would you happen to have a map or be willing to lead me to the path out of these woods?" she asked.

"I don't think you want to leave," he said with something odd in his voice.

Getting nervous she asked why he sat there a moment waiting to answer seems to enjoy making her uncomfortable.

"It's going to be dark soon and even if I were to show you the path you might get lost again, or worse," he added with a stern expression.

"Dark?" Catrina asked for they'd just eaten lunch it couldn't be later than 1 p.m.

"Yes," he answered.

She jumped up and swung open the front door gasping when her eyes were met with darkness outside. Confused and afraid she turned around a question on her lips, he wasn't there.

Calling his name she searched the house not finding him and started to panic. Why would he have just left her alone in his house she was a stranger, she decided not to stay around and searched for a bag and some food to carry meaning to pay him back later.

She left a note thanking him again and set out to find a place to sleep not wanting to sleep in his house for he made her feel very uncomfortable despite his generosity.

Baffled at the fact that it was night time she walked blindly into the woods again trying to be quiet, hearing footsteps she stopped hearing her heartbeat like a bongo drum sure whoever it was out there could listen to it as well. "Catty," a voice whispered close to her.

Crying out she ran tripping in the near pitch black of the forest; she slowed to a stop gasping for breath and sobbing when she realized that it had been her husband's voice.

Impossible, she thought to strain to look into the darkness not daring to call out in case he was out there playing a cruel joke.

She pauses for a moment, while she caught her thought while wishing that was her husband, but started to run back toward the cabin. As she approaches the cabin, she stopped in her track and heard Raven, talking to someone.

She was walking closer to the window pane, in which was cover with a glair of dirt. She eases over to the window, she put her hand in the pain, and rub it slightly, and therein behold, she saw her husband, and he was sitting there with chains on his neck.

Before she left to go to venture out to find her husband, she sent the detective a letter, in which she was going to find out what happened to her husband. While the inspector was concerned, he calls his best friend, in which was a retired park ranger. The park ranger came to his office to hear what he had to say.

The first victim that went lost in the woods was a mad scientist, that was the band for his profession, because he thinks that the mind can be controlled by taking the will of a human and turning it into a beast. So the ranger processed his data and decided to find out what is going on in this parts of the wood.

The Sherriff dept gave him two officers to go with him. Meanwhile, Catrina was shocked, and disparate to see that he in-deed still alive, and he is in the place where she just left. As she observes the moment of to two, she looks over slightly within the cabin, and she saw a hole in the floor, in which look like a cave beneath.

As she presses her ear closer to the cold window pane, she heard him saying to Stephen, 'that lady that is here, I need you to find her, and bring her back to me." "She will tell the thing to you, to make you think that you know her, but you don't, you only do what I command you to do.

About that time, he pulls out an electrical stick, and shock him until he passed out. As the ranger and his two officers arrive at the interest where she went in, the Rangers just stops and focus on what is the difference about the layout, and while the wind started to blow, every so breezy, he realizes that the pathway is cover with leaves, and puddle of water.

He looks over at the officers and said gen-
tlemen, this will impossiable and harder to
find who we are looking for. So if you want
to turn back now, with a stern look on his
face,' 'here is the time''. They looked at
each other and said, ''Sir let fine who we
came here for'', and they went in.

Catrina was very scared about what just
took place, and did not know what her
next move to do was. So she had to leave
from there to get help, and she made
haste back through the woods to get help.
It was early that morning during these
events.

took place, but when she got 200 hun-
dred yards deeper, she heard Raven say,"
You can run, but there is nowhere to go."
In her mind now, that the man she loves, is
not the one she saw.

The Ranger and the officer were headed
deep in the forest, their boots were getting soak, and they were getting bitten my
bugs and ants. The Ranger told the others,
'hold here, with a client look upon his face.
He bent down, and there was a piece of a
torn bag, with wet camping supply all
around it. One of the officer went to step
further, and found a camping knife,

 That she lost. The woods were silent, you
could not hear a bird whispering in the el-
ements, and the Ranger said to the others,
'this way.

When Catrina got further into the forest, she stops, to catch her breath. In her mind, while she pauses for thoughts, she could remember back when they had their first child together. While lying in the bed,

 Delivering their first child, Stephen was coaching her how to push, and he said to her," baby, you must get your breathing alone with your pushing, don't worry I will never leave your side". About that time, she heard a branch break about 20 feet behind her. With fright, she just stood still.

As she glazes her eyes, ever so slightly, she saw Stephen, and he was like he was in a trance. She wanted to call out his name to him, but she did not want to hear the monster within. She stayed still until he left ahead of her path. She said to

 Herself, I love Stephen, but the only way I can save him, is to go back and find out what Ravan did to him, and she went back another route. She notices that Stephen was wearing a collar around his neck, with two light flashing, and in her mind, it must be a G.P.S. system, so she hasted back to the cabin, to find answers.

While Stephen was looking for the lady in the woods, in which in his mind, honestly did not know who she was, he came upon a boundary line, and he stops.

He was aware that his collar would give him a great shock, so he turned around and headed back. He looks like a wild man that had not taken a bath in years, and the clothes he was wearing, just ragged. In the years there, Ravan, who is the Mad Scientist that went mission, Stephen, do have some weak memory left every now and them.

As the Ranger and the officers came to a big tree, which looks like it has been occupied by someone last night, he knew that she must have been here. He got closer to the tree, to see, what looks like blond hair, and some branches that have been broken for cover. They call out for her, but there was no answer, just echo in the midst, so they pick up haste deeper in the woods.

Catrina started to head in a different direction, climbing over a stump and slipping into the wet mud. When she finally saw the cabin through the damp trees, the ground gave in under her feet. She feels about 10feet, her eyes got dimmer, and darkness came upon her. While the noise came from the fall, in her mind, she was at a 5-star hotel, and she remembers that there were roses peddle forming a path to a hot bath.

The light was dim, and a glass of wine was waiting for it first drink. She got in the temp water, in which was perfect to a passion of her skin to behold, Stephen was at her feet and his smooth touch to her feet, just put her in a trance.

The song he had playing for her, was "A Kiss from A Rose" by Seal, and about that time, Stephen got into the tub with her.

 It was getting dark in the wet forest, and the Ranger and his officer, camp out by a fire.

One of the officers went to get some woods to make a fire, while the others were putting up the tent. When the officer that went to get some wood for the fire, he heard a noise like someone was walking through the path. He calls on the other, and they came with haste.

The Ranger, who, John, came to the area where the sound was, and he saw what to be a print of a man boot. One of the officers asks," what do you think that it could have been", and he replied, "so we are not alone, and I think there is more to meet the eyes here." You must be alert at all time, and they went back to their camp.

The fire was blazing hot while they were sitting around the fire. The Ranger pull out a map of the layer. He told them," men about 300 yards ahead, there is an old cabin on the hill, if it still there, tomorrow, we will camp there.

When Catrina awakening, she hears voices, as her eyes begin to see closely, she saw Stephen setting in a corner, looking directly at her, while Raven was going through her small wallet she had in her possession. He notices that she was awakening and he said to her," why did you try to run, what did you think that, I was going to let you go, with a deranged look on his face.

She said to him," why are you doing this to my husband, he is a good man, with tears in her eyes. While they were talking, Stephen began to move his head like he does know her but cannot connect the emotion. When Ravan saw his moment, he shocks Stephen to sleep.

As he drifts away, Stephen memories start to kick in. He saw what appeared to be a wedding; he was walking down the aisle, looking at his bride, in his thought, she was beautiful, but faint from a distance. The song was playing, 'I Will Always love You, by Whitney Houston, He was smiling with happiness, and the closer he got to her face appearing started to become more manifest, and all of a curtain, he woke up.

Catrina asks Raven," what are you going to do with us, and he said," you to will be my subject or torcher, and how I can turn human emotion into animals. The first subject I experienced on, could not take the pain,' then she asks,' what happy to that subject, with fright, and the told her," you were eating that subject, when you arrive here,' Catrina.

She started to throw up like a rushing out-
pour. Raven said, as he walks to her carefully,' and if you can't evolve to what I
want you to be, while smelling her hair,
you will be next. Raven then gave her a
neck bracket and told her to put it on, and
she did. The next morning, the Ranger and
his officers called back to headquarter but
had no signal, and he told the other two,"
this may be a search and rescue, and yes
we don't know who, are what we dealing
with, do you want to go back before any
further, what say you.

As they gather among themselves, the officer felt that they made need some backup and said to them," Sir, if make asks, why don't I got back and bring some help. The foot track was too fresh, and there is more to meet the eye here, and the Ranger said,' go and bring help, and he went. As the two got further in the woods, they saw smoke coming from the cabin, and the Ranger was shocked that anyone was still living there, so he told the other officer, we may have a crime out here be a lark.

When Raven put the collar on her, she looks over at Stephen and said with a loud voice, with tears in her eyes and said," Stephen I love you, this is your wife, 'Catrina, please here he within. We have two children, their name is William and Michael, they miss you a bunch, "you got to remember, 'please.

About that time, Raven had enough of the conversation and picked up the shock stick and put in on her back and before he could give her the full charge, his alarm went off he set in the wood. He stops with haste and picks up his gun to see what was going on. He asks Catrina,' bitch who are with you, with a loud voice, and with tears in her eyes, she said no dam one sir.

He walk over to her and said,'' I am going to ask you one more time, he pulls back the handle on his gun, in which at that time sounds like a tick from a clock, and she said,'' I left a note to the Sheriff sir,'' that really all, please don't kill me. He told her, if I come back and you move an inch I will kill you both, that will be enough food for two years, and left.

The Ranger notice the wire that was at his feet, and he told the officer, "stop" with a concerned look on his face, and he bent down and followed the wire to a monitor. He pulled his gun out of its holster and told the officer" this has just become a problem that may get dangerous, keep your eyes open.

Stephen regain his strength and begin to just look at Catrina, while her head was hanging low with disbelief and all of a curtain, he said " Ca-tuna', with an exalted voice, she raises her head with tears just flowing tears and told him," baby yes, "yes this is me.

'' About that time she moves her chair with a forward motion and grab his hand and said,'' do you remember,'' do you remember when we in in high school, the final game that you threw three touchdowns to win the game baby, with a smile on her face', and do you know what you told me that night Stephen', and his eyes were just going back and forward, and he looked up and said,' I, I, I love you, and he came out of the trance he was in.

When Raven got to the area that the alarm went off, he heard two men talking, and he laid down for the ambush, two seconds later, you heard two shouts. When the shots roared out through the forest, Catrina said to Stephen,'' baby we must get out of here now.

Raven walks up to the bodies and look at the Ranger and said,' early this morning, you sent an officer to get help, with a pose look on his face,' but let me

tell you he only died in a second, as the dying Ranger just look on, with little life left in him, and Raven look down at him and said,'' I wish he were a little bit leaner'', and he died.

When Catrina and Stephen heard the shoots, they were trying to get themselves lose, before Raven gets back. So Catrina looks over to her left, and right on the table, was a knife. She rolls her chair over toward the table, but her hand could not reach that for.

She started to panic, and she remembers. One day while riding their boat in the lake, it was a beautiful day, the family was having fun, and her son, William was swimming, and it was time to go before it gets dark.

William was trying to get up on the boat side, and he said to them," mama, with one hand on the bow," I can't pull myself up,"' and she told him," use every part of your body to get in," you can do this.' So William, while having on hand on the ladder, started to swing his body back and forward until he got in. When she realized that Raven was headed back to kill them, she took her feet, and pull the knife toward her.

Raven hid the three bodies close by and headed with haste to the cabin. He knew that someone heard the shots, so in his mind, he must kill off the rest and move up river. He got to the cabin, pull out the rifle, and kick the door in. Without haste, he ran into the cabin, but they were gone.

He cried out with a loud voice," I will find you, and I will kill you all. "There is no one coming for you. 'Running with fear, they begin to travel through the forest like there life is in danger, and Stephen was trying to get his mobility back.

He told her," baby, when I got lost a while back before he found me, I have a secret place I know, we can hide there until I get stronger, and they went. While waiting on the progress of what is going on with the search of the victims, the Investigator call in the FBI, to handle this case.

The Agents in charge that arrive the next day gains Agent Quinney. The Agents have been with the Bureau for 29 years, and he knew about the missing bodies early, but he let the County authority handle the case.

When they got in town, they went to the Investigator and " I need all you got on this whole case and don't leave anything out, and the Investigator said, 'yes sir.

Meanwhile, Catrina and Stephen arrive at the place where he stayed back then when he got lost in time past. When she entered the cave, she was so tired, she just becomes and went straight to sleep on the damp floor. Stephen went and got some branches to cover their track, and to put under her body to get out of the cold ground.

After building a small fire, he realizes that she has been looking for him the whole time, and he fell to his knee and just started to cry and put his hand on top of a head and told her," I will never leave your side, and we will see ours

children. Somehow after that, she smiles within. Now Stephen had lived with Raven for Over three years, and vaguely, he knew his every move, so while She was asleep, he decided to come up with a plan, to kill him for good.

Raven knew that he had to kill them, to
continue his madness somewhere else, so
he pours gas all around the cabin, and lite
it ablaze, then he jumps in his wooden
boat and sails in the direction where they
fled.

The next day, Agent Quinney and his men,
gather outside the woods for a short brief-
ing, and he said to them, these woods
have secrets that no one can figure out.

 To my understanding, we may be looking for a serial killer. We have six peoples in all, that came up missing, and one of them must be the killer in these wood. Most of us are train to survive out here, but there is more at work here. About that time, here come the helicopter flying overhead.

So they enter the woods. The sun was shining, and the dew on the trees was dry-ing up. As they approach the first seen, it looks like someone has camped out here. They collect all of the evidence and push on. All of a curtain, one of the Agent said," Agent Quinney you need to come over here and see this, and with haste he did. The first officer that left to get help was dismembered like someone was packing the parts to be preserved.

He was stuff in a Beaver Dam, because the temperature was a chilly freeze, and the Agents look on with concern about that time, the helicopter Agents call to Agents Quinney and said," about 500 yards to your 12 o clock, there is a small cabin burning, and I can't tell what because of the trees, so I am going back to the station. They don't have air cover no more.

Meanwhile, in the cave, Catrina finally awoke. She jumps up with fear, thinking all this has been a dream, but when she laid her eyes on Stephen, she just embraces him like that first meet. So Stephen told her,''; baby, about the time someone finds us, we can't outrun him, because he has been living in theses wood for years. Now to our advantage, I have been his puppet for quite some time. He love for someone to act like they are an animal to him, but if not, he will eat us, like he will consume our soul

She looked at him and said, with concern, what is there for us to do, and he said, 'to fill powerful, you have to give more than they ask for.

When the Agents got to the cabin The Agents said," who in there right mind will do this to these officers, with his hand over his face, because of the smell of burnt flesh, what they saw was horrible Raven has taken the bodies of the Ranger, and the officer, put them in the cabin, tied them up to a chair, like they were looking at one another.

Then another Agent said," you need to come and see this, he went with haste. There was an old deep freezer in the back of the cabin, sitting about 20 feet behind the cabin, and when he looks inside, there were about two more dead bodies, and the smell was horrific.

Stephen pauses for a minute` and said to Catrina," baby while I was captured, there was another voice that I heard while I was captured, but when you arrive, the voice was no more.

There are six Agents in all that handle this case. So Agents Quinney said to them," we all must break up into three groups, and the layout here, we should meet here by late evening, as he looks at the dead bodies with concern.

The Agent, pull out his computer and started to look up the peoples that went missing years ago in the forest. As he such thought the information, the first person he came across, Was Dr. Ravan Mullins. So he begins to research the history on this Doctor. Dr. Mullins, was a mad scientist that think human behavior could be control through electric shock, and feeding it flesh will help the process, and at that moment, he knew who he was looking for.

Stephen told Catrina, as I started to get better as time goes on, I can also remember, the name David for some reason, it like he was helping us do some of this destruction. Raven was hot on their track, and he knew where they were going. About the time, he was headed in their direction, there was a noise, like a phone ring, and by that time, he pulled out a satellite phone and said, " what is going on." The End."

Part 2

When Raven answer the phone, while breathing uncon-
trollably, he just listens for a few second and hung the
phone up. The expression on his face looks like he
heard, that there were others in the woods also.

Since the woods were so dense, his tracking ability
came to a halt, and he camps out at the edge of the
dam, waiting on a movement.

The Investigator, Agents Quinney, went back to the
town, you gather more Information about the scientist,
Dr. Raven.

So he calls his agents in, and ask them about what do
they think is going on, in the woods. One of them said,
'Sir, the woods have too many secrets in it."

He said," how can this type of horrible situation go on,
without any worries from the community, with a confused
look on his face.

Agents Quinney said," I want you two, to monitor the activities they roll on outside of the woods, why the rest of
us, question the town folk.

Stephen and Carina were still in the cave, they finally
got back to normal, and they were hungry.

Stephen knew that they were being hunted, but the rush
from his stomach was too much to control, and so he
said to her." baby, I will be back, I am going to get some
food from the creek bed, do not leave this cave,

with a sincere look on his face, and she said to him,"
whatever you do, I will come with you, I will never leave
your side while we are in here.

While it was early that morning, the sun has just come up, the ground was still wet, and the woods were alive with many noises.

Raven got up and started the hunt back for them. He was looking around every corner, and while the dense fog rose from his mouth.

Then It hit him, where they were, and he took off with haste. While Stephen was bending down at the creek getting some trust liquid for them to drink, Catrina was in her though.

She remembers the time they were going on a horse ride, through the plain.

How Stephen was looking with his tight blue jeans on and they sexy body that she ever saw. As he was walking toward her, he put his hand out and said, 'grab my hand baby, and he put her a top of the horseback.

While still in a trance, Stephen was saying, "baby, 'baby, then she came out of the moments and said," Sorry baby, I just had a moment."

Now Stephen is aware that Raven is coming after them, but he is trying to think, while the hungry was taking over.

All of a sudeen, while standing by the creek bank, here comes a dead fish, floating down the stream.

The smell was horrible, but the illusion of meats that had been unjust, and Stephen Jump into the water and pick it up.

Without any hesitation, they consume it like wolves on prey. As they Creek approaching the cave, there was a sound inside, like an animal looking for food, so they easy back and bent down by a tree stump, holding the steam within, while the frost was freezing there the bone.

An about that time, they saw a gun, sitting at the entrance of the cave, and then arose Raven, the facial look like a killer.

So they just sit there, in the cold, while he moves from the entrance. Stephen was thinking, about that time, they were playing hide see, with their children, but the thought did not last long, because Carina said," he has gone back in, so let go, and they did.

The Agents was trying to figure what is more to this situation than meets the eye. So he looks up the Scientist name, started to figure out more about him.

One something he notices, that he is from this small town here, and his father was well known Minister.

His father leading groups of followers and got release from the church for some misbehavior.

About that time, he came across a picture, of some members of a particular organization, and there was His father, himself, and three others, but he could not put this together with the point at hand.

So he calls the others agents, and ask them about ant activities they may have occurred, and they said," it quite here boss" and they hung up the phone.

As he was leaving the office in the small town, he notices that most of the peoples there look as looking at him strangely, as they knew going on out there.

So he got in his car, with a copy of the picture in his hand, and headed to the Sheriff station.

After his arrival, he was looking at some strange symbols, which seem to be everywhere, but he did not pay it attentions.

While Raven was waiting for them to get back, Catrina said to Stephen, 'we can't keep this up and expect to find our way out", with a frightening voice,

and he said to here," do you remember what I told you the other day," in order to have killed a lion, you must know how to out swift him, so here is our plan.

While at the sheriff office, Agents Quinney is looking for some answer quickly, if there any chance of finding out, is anyone still alive.

So the sheriff came in and said," are you having luck with what in the hell is going on in those woods.

I want to tell you a story about time past. My deputy will not go in there, because of the secrets. This country has lost plenty of peoples in time prior.

You see, the lost we just suffer, it like someone is who-ever knows when someone enters the woods.

So the Agent asks, who jurisdiction it is the monitor the wood out there, " and he said the Rangers, of which are probably dead, and the city police, but they don't even patrol the area with a concerned look on his face.

So the agents pull out a picture of the Scientist, and his father in the picture, and ask the Sheriff, can you tell me about any of these peoples her.

While looking at the Sheriff and he said," when I was young, his father was the town minister, and he had power here, but when the church rebel against him, and he had a few of the town folk went with him.

Have you paid attention to some of the symbols her"
and he said to him," the ones that look like a Raven,"
with concern, and he said yes, well that the symbol of
the Society, and at one time people feared them?

It is said in time past; they use to sacrifice others that
drifters in the town., in which I think they are out of mine,
with a smile on his face, so peoples don't ask the question.

The Agent ask him, does he have any family left here,
and, 'is there someone else I can talk to about what in
the hell is going on,

we have lost officers in these woods in the last few
days, Sheriff, with a loud voice, while standing up, and
said," look here Sir, when others peoples start to come
around here, this is what happens, we don't know a clue
where they went.

After tension had gone down, the agent show him the picture he had in his possession, and ask him," is anyone here you know" and the Sheriff said," these two are dead, but I don't even know who the other person is, but if you go to the town Historian, he may help you, and he left.

Meanwhile in the woods, Raven was waiting for them to come back. Then he got a call and he said, 'listing, we must clean up this mess once and for all, you all know what to do, I got unfinished business here, I will contact you when it finishes, and by the way, 'kill them all, and he hung up the phone.

After hearing the loud conversation, they went back, and Stephen said," are you ready to go hunting", and she said," I thought you never ask."

About that time, Catrina stood down the hill for Raven and when she saw him, "she said to him, while having a spear in her hand, 'No more running" while frost was coming out of her mouth.

All of a curtain went out of the cave, looking around to see where the voice came from and she said, 'you fool,

it time to hunt, and before he could raise his raffle to
shoot, she took out with haste.

With a loud voice, he said to her," today you all will die"
and he took off after them.

The Agent went to the courthouse to talk to the Historian, and he came up to his office and said," Sir I got
some question for you, and I need some answer with a
stern look on his face, and he told him to sit down.

The agent told the historian," do you know about "Society", the all of a sudden, the historian said," you need the
leave this place, before you and the rest of your officer
get killed.

The agents were very shock of what he said and told
him," you don't have to be afraid, I know talking and
been in this town, and the woods are connected with
missing and the killing of other.

Then he told the Agents, with a fright look on his face,
'have anyone follow you over here, and the Agent said,
'no, why" and he told him," this town has secrets also,
some of the people here still believe in the minister
teaching, and peoples just don't talk, but they just do.

So the Agents took the picture out of his pocket, and
show it to the Historian and said,"I know this is the late
minister, and these two are dead, and the man in the
woods, is Raven, but who is this one here.

About that time when he was about to tell him, there
was shot from a high powered rifle, and he died, and a
truck speed off.

The agents took cover, and call his agents who were at the edges of the woods. The agent's phone rang, but there was no answer.

In his mind, he knows by know; the woods is a breeding fields for death.

Catrina was running down the muddy path, and while Raven was catching up, and about the time, she fell on the forest floor, and when Raven got in range to take the shot, he fell into the trap that Stephen had set for him, and he knocked himself unconscious.

Stephen and Catrina walk toward the hole very slowly, pick up the gun, and they saw Raven, laying there, with no movement.

Then Stephen pulls him out of the hole, drug him back to the cave, and tried him up.

When Raven came to, he realizes that he was tired up. He glazed around and said to them," what do you think this will do for your cause."

You all don't understand the society; we will never die, looking directly into their eyes.

Then Stephen said," for years you torcher me, and I have seen you not only kill peoples, but you also eat them.

You have stripped my humanity out of me, and you hurt my love in front of my face.

So I decided to grant you a slow and painful death. About that time, here comes Catrina, looking like a wild woman, she put the shock collar around his neck and turn the power up high as it can go, and made him pass out.

They both were sitting down, looking at Raven, and he said to her," baby I can go back to civilization, because of the events took here in these woods, but I want you to go back,

you know why I can't go back with you baby, with tears in his eyes," she baby I don't care what you did, and who you became, but what I do know, I love you" and she huge him with fear.

The agents don't know what to do, but he knew, he had to get out of that town soon.

So he made his way to the police dept and told the Chief, I need to call the F.B.I. And the Chief said," all the tower are down Sir', but we can help you.

The agents told them, we must get to the woods now, we need to warn them that we are dealing with a cult in the woods, with an exhausted voice, and they left with haste.

So about that time, while they were driving the road, the agents said," is there a tower by for me to get some connection, and the Chief said," one mile before we get to your agents, you should be able to make the call to the Agency.

Stephen said to her," I know a way out Catrina, but we will have to leave soon, today is the first moon, and they will be here for worship, but before you go, I must do this, so you would know this man knows more.

He set Raven upward, and start saying different words, and after that, he looks over at Catrina and said, 'his

power will be mind, and slit his neck from one end to
other.

When the blood starts powering, he put a cup and
caught about a cup full and drink it until it was no more.

She was afraid of what took place, but now she knows
the man she craves for, is no more, and said," this is not
your fault, but this demon here, but I want the man inside to know I will never stop loving him, with tears in
her eyes and then he took her by the hand and left.

**It was late evening when they got to the forest, and he
saw the two-cars that the agents were driving.**

He jumps out of the Chief car with his gun in his hand,
and the Chief got out also.

There was no moment in the cars, so the Agent told the
Chief, "you go that way, " and he did. About that time, he
looks over in the car, the two officers were shot to death,
and when he got over to the next car, where the other

two was, one was shot in the head, and the other was missing.

About that time; the agent was holding his head down, the and things started to get clear. When was asking everybody in town about the missing and killing, some of them had a mark on their middle finger, like the symbols around town.

While his head was still down, he remembers, when he ate at the dinner, the symbols were there, and when the Sheriff stood up, he had the mark too.

Then he remembers, Raven last name was Mullins, and the person that was in the picture was the Chief, his little brother and about that time, with his hand on the trigger he swirls and before he got the shot-off, the Chief shot him in the head.

Later that even, when they came to the edge of the patch, she finally saw the road she needed to get home.

She said to him,' do you remember our first kiss Stephen" and he said, "until the day I die love", and she stood there, her face was pale, her hair was pull back, and look heartbroken.

He said to her, tell my sons that I was found dead, and I want you to leave here, and never come back.

She took twenty paces out of the woods, and he was gone, and she left and went home.

Three days past and Catrina had pack up and got into U-Haul and pull off. While she was driving down the road, she begin to think about the time, last time they made love.

Stephen came into the house on a long day of work, and
when he opens the door, there was a letter on the coffee
table said ," I want to play a game with you tonight, but
first, go to the bathroom, and get into the hot bath I
made for you, and he did.

When he finished up, he looks over to his right, while the
soap suds was pouring down her back like rain, there
was another letter that said, when you get out, come
into the bedroom, but don't cut on the light.

When he enters, with only a towel on, he heard voice
said," Stephen I have been wait on you for a while, with
a passion in her voice,

Then she told him," you can turn on the light now, and
what he saw, she was laying in the bed, with only stock-
ing on, and a seductive look on her face.

He easy into the bed, spread her legs wide open, and
her eyes went back into her head.

He was stroking her body so deeply, until about that
time, here she cumm, then from the blow of a car horn,
she came back to reality.

She pulls over to the woods for the very last time, all the crime scene was going, so she blew her horn twice, and waited for him to show, but he did not come, so she started to pull off,

And all of a sudden this figure appears deep in the woods, and she smiles, and he turns around deep in the woods he went. Then she pulled off with a smile and was never seen anymore.

Later that night, you could see a fire deep in the woods, and the road at the entrance was full of cars.

As the crowd gathers in the circle, chanting world from a book in their hand, here comes a figure in a black hood, and it was Stephen.

When he killed Raven, he gains his strength and power, in his mind, The Chief pulls off his hood, and said," tonight we will feast off a warmblood, and about that time, they brought out the agent that they did not kill.

They put him on an altar, while his tongue was cut out, and the Sheriff, took his hood off, and in his hand was a dagger, then he shouted out some words from an old book, and put the blade into his chest, and death took the agents.

Two years have passed, and the woods keep its secrets, and Catrina has put the memories to rest.

She doesn't have the passion through anymore, but now and then, she misses Stephen.

The next day near the park, a family that was headed on vacation, their car broke down at the entrance of the

woods, and the husband said," you all stay here, I am
going to find help, The End.